Emergency Sun:

How to Build a Portable Solar Power Supply for Smart Phones, GPS, Cameras and Other Electronics using Rechargeable AA Batteries, Design, Parts, and Procedures

by Christopher Kinkaid

Published by Solardyne, LLC
Portland, Oregon

ISBN-13: 978-1523616770
ISBN-10: 1523616776

Table of Contents

Preface

"Power, power, everywhere - nor any drop to drink"

In times of Emergencies, electrical grid power is the first to go out. Blackouts, Floods, Hurricanes, Tornadoes, Earthquakes, Land Slides, the natural world can unleash catastrophes which leave people without basic services, and especially sources of electricity.

Solar energy is an excellent source of power for charging batteries especially in emergencies and prolonged blackouts. Available in some degree everyday and everywhere on the planet, Solar Energy can be a practical source of power for use anytime day or night for powering Smart Phones, GPS, Cameras, and other Consumer Electronics in an emergency.

This book describes how to build a solar power supply and system which is portable, rugged and sturdy in construction, highly efficient at battery charging, and produces DC power through a Universal 12 VDC "cigarette style" plug or USB output day or night on demand to power electronics for Outdoor use and during Natural Disasters.

A personal rugged power utility, the "Emergency Sun" is a portable solar power system which charges internal Rechargeable AA batteries to provide power at anytime and nearly anyplace by simply plugging-in.

Available in most hardware stores around the world, the Rechargeable AA battery format as the storage technology of choice provides a least-toxic battery technology, universally available, safe to fly, ship, and travel, and provides up to 600 charge and discharge cycles per typical battery set - allowing daily use for about a year and a half on one set of batteries. The Solar Panel has an estimated life of over 25 years for a long useful life.

The Emergency Sun, Patent Pending, design described in this book can be built using the Emergency Sun Kit which contains all of the parts you'll need, or you can source the components on your own. A complete parts list and tool list is provided and complete procedures for building your own portable solar power supply.

A portable solar power supply is useful for Campers, Outdoor trips, RVs, Boats, Hiking and most especially during Emergencies, Black-outs, resulting from Man-made and Natural Disasters. In each of these occasions the needs are the same: power on demand to run vital electronics such as your Smart Phone regardless of time of day or immediate situation.

Natural disasters provide an especially challenging scenario for designing and building solar power supplies. Physical shock, vibration, moisture, temperature changes and extremes all conspire to create an environment which is tough for electronics in general, and power supplies in particular. The "Emergency Sun" is designed to be a useful power supply in a real world. The 1500 mAh's of energy stored by the internal battery bank, and the thermodynamic balance of the Solar Panel charging voltage and amperage with Twelve (12) AA REchargeable batteries give the "Emergency Sun" a powerful kick, and provides useful power on demand, and energy to charge and power your electronics.

In our digital age the use of Smart Phones, Cameras, and other micro-devices drives a large portable battery market. Unfortunately, many throw away batteries are highly toxic and present a danger to our natural environment from toxic leaching as the batteries break down.

The Emergency Sun uses non-toxic Rechargeable AA batteries which present the least toxicity in our battery choices. Twelve (12) Rechargeable AA batteries will typically charge and discharge in full over 600 times providing many months of daily use possible.

Build your own Solar Power Supply using the Seventeen (17) steps described in this book to provide power for Outdoor use and to provide portable emergency power for electronics during a Natural Disaster.

About This Book

Emergency Sun is written as a Step-by-Step guide for building a rugged and Portable Solar Power Supply for powering consumer electronics indoors or outdoors, day or night. In addition to construction and soldering instructions a complete parts list is included which you can source independently for personal use,, or use the Emergency Sun Kit which contains all of the parts you'll need to complete a finished Portable Solar Power Plant.

Solar Power systems must meet a wide criteria of standards and endure many stresses in the field. A portable power system must perform as a moisture barrier, a mechanical shock barrier, and an electrical barrier to keep your batteries and control circuit dry, mechanically secure, and reliable when needed in an emergency, or any other time of desired power on demand.

To construct the Emergency Sun Kit or from parts you source independently, you'll use mechanical, electrical and aesthetic skills to prepare your wires and electrical components. Mechanically, you'll layer the Master Gasket, assembly your Back Panels and mount your electrical components into a circuit and Test all materials and circuits for proper function. The final assembly brings all of these elements into a Solar Power supply which is portable and ready to charge in the Sun, and power

12 VDC electrical devices, or any Smart Phone, GPS, Camera or other electrical device via a 12 VDC/USB converter giving USB output.

A Quick Guide is offered at the end of the book which lists Tools, Parts, and Procedures to give those ready to begin a starting point.

The first Two Chapters of this book deal with the subject of Solar Energy, and Solar Power Systems in general. If you wish to go directly into the discussion of construction go directly to **Chapter Three**.

Chapter One - Solar Energy as a Power Source describes the Solar Resource and how photovoltaic solar cells and panels are constructed to deliver Direct Current power. Solar energy contains a respectable power density of 1,000 watts per Square Meter at peak and presents a substantial power and energy resource for industrial purposes, including powering remote electronics.

Chapter Two - Solar Power Systems goes into the design aspects of approaching power systems for powering electrical loads from remote electronics including but not limited to Cameras, GPS, Smart Phones, Tablets, Laptops and other electronics to larger electrical loads including off grid Home Power, Water Pumping, and Water UV Treatment systems. Describes important aspects in considering and calculating the size of a give Solar

PV Panel, Battery Bank, and output devices for solar power systems in general.

Chapter Three - Emergency Sun Parts and Procedures - Begins the discussion of Emergency Sun as a power system and describes the Tools, Parts, and Procedures you can use to build an Emergency Sun power system from your own sourced parts, or from the formal Emergency Sun Kit.

Chapter Four - Preparing your Solar Panel and Wires. The Solar PV Panel described is rated at 5 watts and made from a photovoltaic laminate housed with an aluminum frame. This sturdy construction becomes the frame and backbone and structural strength for the portable solar power supply. In this chapter you'll cut and strip your Solar Panel wires in preparation for building and interconnecting your control circuit.

Chapter Five - Prepare your Wires for Soldering. This chapter describes the cutting and wire stripping of various electrical parts and other preparations which you'll use in later steps when you build the final control circuit. Building these components from parts and completing electrical and mechanical testing provides a reliable and safe interconnection allowing your Personal Power Plant to work with high efficiency and performance.

Chapter Six - Building the Control Circuit. This chapter covers soldering and Heat-Shrink use building your Control Circuit which integrates the Solar Power inputs, the Battery Bank wired in Series, and the Power Output Leads. The heart of the Power Plant, the Control Circuit protects the batteries and insures that no electrical back-flow or leakage occurs once your Solar Panel is charging your Batteries. The Power Control Circuit is wired and tested in this chapter with specific soldering procedures and protocol.

Chapter Seven - Electrical Testing. This chapter describes how to measure and test your power plant to be sure your Battery Bank is operating at proper voltage. In this section we'll test the Solar Panel for proper function and with the Control circuit built we can test the Solar Battery charging function. The final tests will verify that the Power Output plug is operating at proper Voltage. Testing includes electrical and mechanical strain tests at each step to insure each solder and connection you build is secure and safe.

Chapter Eight - Mounting the Circuit. This chapter describes how, after testing, you can mount your Electrical Control Circuit to provide electrical isolation as well as mechanical shock protection. Proper mounting of the control circuit using provided Sticky Pads provide an advanced way of protecting the physical and electrical circuit.

Chapter Nine - Building Back Panels. Once the Control Circuits have been built, tested and mounted the next step is to build your Back Panels.

The main Back Panel provides protection for the internal electronics while allowing access to the Battery Bank (for loading your batteries), and access to the Power Output Plug for normal use. The Back Panels use industrial Velcro as the attachment mechanism. Velcro makes a surprisingly effective barrier yet is relatively easy to remove and replace the Back Access Panels. In this step you'll measure, cut, and apply the two Velcro surfaces to the Back Panels and Panel Covers providing a convenient and effective Access Panel method.

Chapter Ten - Mounting the Hardware and Final Assembly. In this Chapter we begin the final assembly by mounting the Main laminating Gasket which integrates the Solar Panel and control electronics with the Back Panels. Prior to this step all electrical testing has been completed before sealing up the unit. Steps and procedures are pictured which describe how to assemble and construct your Back Panel assemblies. Final Assembly results in a final test by loading up your twelve (12) Rechargeable AA batteries and powering your Smart Phone or other devices directly from the 12 VDC Output plug or USB converter.

Chapter Eleven - Using your Emergency Sun - As a power plant your Emergency Sun is now available to charge up and provide power anytime you need it. This chapter describes how to load your batteries and access your Power output plug. As a solid state device, the Emergency Sun has no moving parts, completely silent and powers your electronics with no toxicity or emissions. As a power supply the Emergency Sun is designed to be rugged and survivable and be flexible to provide power for your Outdoor camping, or when in your car for emergencies. The Emergency Sun

Chapter Twelve - **Quick Guide** - **Tools, Parts, Procedures.** This Quick Guide provides a short hand list of the Tools you'll need, a Parts List which you can source independently or through the complete Emergency Sun Kit.

The third part of the **Quick Guide** covers a short-hand description of the recommended Procedures to build your Emergency Sun portable power plant.

Introduction

Natural disasters and other emergencies present a difficult scenario for powering vital communications and electronics especially Smart Phones, Cameras and other digital devices when power goes out.

Solar energy can offer a potent emergency power supply for powering electronics when the grid is down, during blackouts, or when man-made or natural disasters interfere with normal electrical grids. During a natural disaster the normal supply of electricity is the first thing to go out - leaving many people in a vulnerable situation.

Given Murphy's law the moment the power goes out, or a disaster occurs most people find themselves not fully charged up with a limited amount of run-time available on their devices: just when you need it the most.

The national grid is composed of three major regions across the country. Interconnecting multiple power supplies with multiple demands the modern Power Grid is vulnerable to man-made, and natural disasters resulting in Blackouts when most major emergencies or natural disasters occur.

In the Pacific Northwest, for example, one of the major threats to modern civilization is the off-shore Cascadia Subduction Zone, which can potentially cause massive earthquakes devastating the electrical infrastructure over a vast region.

In the midwest, Storms, Floods, Tornadoes, Earthquakes, Blackouts and other natural disasters are experienced and loom in constant threat to the inhabitants. In the North and South-East it's Flooding and Hurricanes. In the West, the threat of wild fires spurred on by record droughts bring a great deal of risk to millions of people regarding the question of powering personal devices such as Smart Phones, GPS, and Cameras during a Natural Disaster.

All of these emergencies require people to respond quickly, often pushing them to move rapidly to avoid further exposure. Most importantly, people's need to communicate during a blackout or emergency where traditional power from the grid is unavailable or compromised becomes a high priority.

In our modern life digital communications are a vital means of reaching people especially in moments of emergency. In an emergency a portable solar power supply with internal batteries can provide energy for electronics and be able to recharge each day through a variety of weather conditions and situations to provide power daily for days, weeks, months and years of daily use.

The Emergency Sun solar power supply is designed to be a rugged, durable platform which provides usable power for Smart Phones, GPS, Cameras and other electronics when traditional power is unavailable. 12 VDC output or USB.

Emergency Sun Patent Pending

This book will guide you Step by Step in building an Emergency Sun power supply suitable for powering your electronics in extreme locations and situations as well as daily use for Camping, RVs, Boating, and other Outdoor use.

The Emergency Sun has Four major elements: optical, electrical, mechanical, and structural.

Your building procedure will address each of these elements and how they integrate into a portable power system. To power your electronics just take out the 12 VDC Plug from the Small Back Panel and Plug in with your 12 VDC adaptor, or a 12 VDC/USB Converter as shown below.

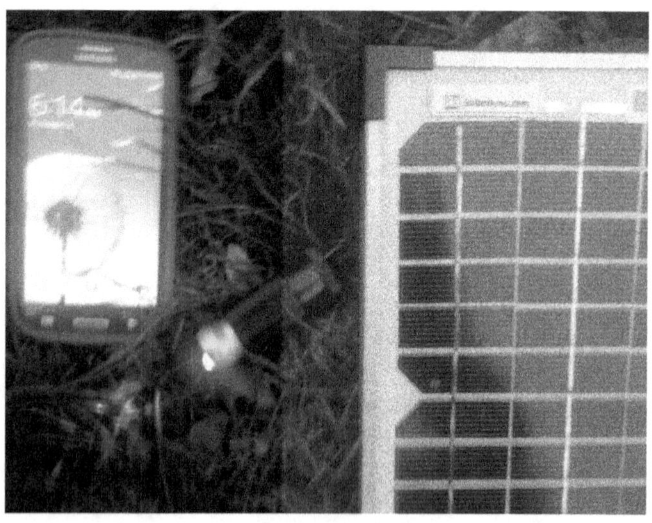

The Emergency Sun Solar Power Supply specifications are:

Height: 11 inches
Width: 7.5 inches
Depth: 2.5 inches
Weight: 2.7 lb. (with internal batteries)

Electrical Rating: 1500 mAh @ 12 VDC
Power Output: 12 VDC or USB with 12V/USB Converter (available in most hardware stores)

Average Charge Time for Internal Battery Bank: 4 Hours Peak Sun or Equivalent

Batteries: Requires Twelve (12) RECHARGEABLE AA batteries

Estimated Life of Solar Panel: 25 Years

The Emergency Sun you'll be building is a solar power supply which charges a set of Internal Rechargeable AA batteries. These internal batteries store solar energy onboard for use anytime day or night which you'll access via the enclosed 12 VDC Plug.

The Emergency Sun is a personal Power Plant you can use to power your Smart Phone, Tablet, Laptop, GPS, cameras or other electronic devices in the field. The power output offers a Universal 12 VDC "cigarette style" plug which you can use with any 12 VDC car plug, or insert a 12 VDC to USB converter (available in most hardware/electronic stores for about $7) to produce a USB output.

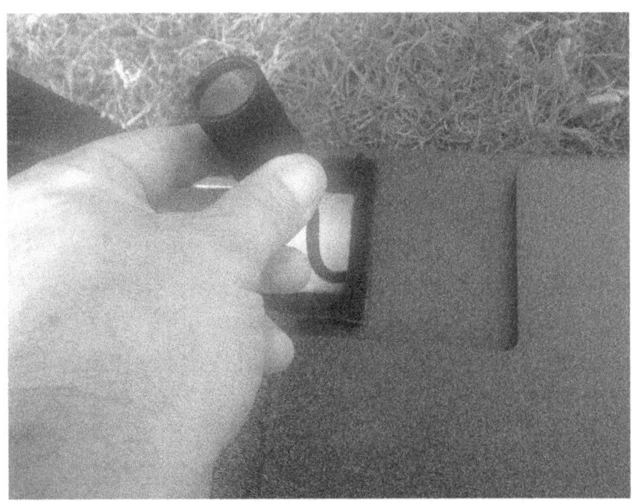

To use your Emergency Sun load Twelve (12) Rechargeable AA batteries into the internal Battery Trays once you've opened the Back Panel. The Large Back Panel houses the Batteries. The Small Back Panel stores the 12 VDC Universal Output Plug.

Once you've loaded the Twelve (12) Rechargeable AA batteries in the back, close up the Back Panels and place the Solar Panel Side into the Sun for Charging.

The average charging time to fill your batteries is approximately 4 hours of Peak Sun or equivalent.

Building the Emergency Sun as Sub-Systems Overview:

The Emergency Sun as a device uses Seven systems which you'll construct and integrate from individual parts and materials into a finished Portable Power Plant.

The First system is the Solar Photovoltaic (PV) Panel which provides the power and energy for the system (Primary Input).

Second, is the Battery Bank which will provide the internal storage of solar energy for use on demand (Storage).

The Third sub-system is the power Output Plug (Output).

Fourth, is the Battery Management System (BMS) which is the control circuit which integrates the Primary Input, with the Battery Storage, with the Output Device plug.

The Fifth system is the Back Panel preparation.

The Sixth system (after electrical and mechanical testing) integrates all of the previous systems by laminating the elements together using the small Gaskets to affix the electrical components to the Solar Panel, and using the Main Gasket, to laminate the Solar Panel to your Back Panel elements producing a complete Emergency Sun.

The Seventh system is testing.

All throughout the building process you'll be testing each solder, each heat-shrink for mechanical and electrical integrity, and test the electrical performance verifying your solders and electrical connections are strong, safe and sturdy for laminating the Back Panel on the final assembly.

Chapter One - Solar Energy as a Power Source

This Chapter is a general discussion of Solar Energy as a Power and Energy Resource using Solar photovoltaic (PV) panels. If you'd like to go directly to building your "Emergency Sun" then jump to **Chapter Three**.

Solar energy is surprising powerful. The energy falling on one square meter is defined under "Standard Test Conditions" as having a power density of 1,000 watts. Digital devices which we use often such as Smart Phones and Cameras often draw less than 2 watts at any given moment. This means a relatively small area of sunlight contains more than enough power to be useful powering modern digital electronics.

From the perspective of devices suitable for emergencies and especially portable devices, portable solar power supplies used for blackouts and times of travel and emergencies require an ability to capture and convert solar energy into electricity, and apply that energy to storage devices such as Rechargeable AA batteries, and finally deliver that energy on demand through a suitable output device.

A Solar Photovoltaic (PV) Panel is a Solid State device which converts Photons into Electrons. The Solar PV Panel does this by interconnecting and laminating for protection a collection of Solar Cells wired internally in Series. Solar Panels are usually made from multiples of 36 solar cells in Series. Larger Solar Panels use 72 solar cells in Series.

The reason for using multiples of 36 solar cells is that each Solar Cell produces about 0.5 Volts. To reach a charging voltage greater than 12 Volts the solar cells must be wired in Series to increase Voltage. The Solar PV Panel achieves this by "hardwiring" an internal Series connection to reach Charging Voltage.

The Current (Amperage) produced by a given solar cell is a function of the wafers Surface Area.

The larger the Solar cell surface area more photons are captured producing more Current. To keep wires as small as possible, and to minimize Power Transmission losses, Solar Cells are wired in Series to build Voltage to about 18 Volts per 36 Cell String wired in Series.

Solar PV Panels range in Power Ratings from several Watts to hundreds of Watts. For the purpose of Emergency Sun you'll build a power system with a 5 watt solar panel. The Power rating of a Solar Panel will tell you how much power at peak sun the panel can produce under load.

To calculate the Energy produced by the Solar PV Panel over the course of the Day depends on your specific location on Earth, and the particular weather at the moment. The National Renewable Energy Laboratory has published a chart of a Location's specific Average Solar Peak-Hour equivalent. This is to say, all of the average energy received at a location on Earth from the sun is added up into a equivalent Peak Sun Hour rating.

For example, a location in Arizona may have 5.5 Solar Peak Hours, with another location like Portland, Oregon having a 3.5 Peak Hour equivalent. Check you location at NREL to get your Peak Sun Hours rating for a given location.

The Energy produced by a Solar Panel will equal it's Power Rating multiplied by the peak Sun-Hours for a given location.

A 5 watt solar Panel in Arizona would be expected to produce an average daily Energy (5 watts x 5.5 hours) of 27.5 Watt-Hours.

If your electrical load draws 27.5 Watts, for example, then you'll have 10 hours of energy to run the load. To calculate your actual Run Time (Hours) divide your Energy Load by your Power Load.

Any electrical device can be powered with a Solar Energy System properly designed.

Our purpose in this project is to build the Emergency Sun type Portable Power Supply for powering your Electronics Outdoors, or during an Emergency.

Chapter Two - Solar Power Systems

This chapter examines general design aspects of solar power systems. If you'd like to move directly to building your Emergency Sun power supply please jump to **Chapter Three**.

Solar power systems are best designed "backward." Which is to say, begin with your electrical demand (the actual electrical load you wish to power) and work the system design backwards to determine the solar and other equipment you'll need to meet that load at some specific location. Solar panels produce Direct Current (DC), batteries charge with Direct Current., and electrical devices run on DC. Being all DC a very efficient power system can be constructed.

Note: Electrical loads are either AC or DC. If you need an AC output then you'll connect an AC Inverter from your Battery Bank and Plug in your AC appliances to the AC inverter. For the scope of this Book we'll examine DC loads for electronics powered from a 12 VDC source including consumer electronics: Smart Phones, Cameras, Tablets, Laptops etc.

The following Design protocol focuses on Power and Energy for digital devices, or any electrical loads which run on DC. This book describes how to build a portable Solar Power Supply for digital electronics including Smart Phones, Cameras, GPS and other devices.

General Design of Solar Power Systems

Every solar power system is composed of Four Fundamental elements. These are Power Input Device (the Solar Panel) a Battery Management Device (the Charge Controller), the Battery Bank (Storage Device), and an Power Output element (proper cabling and Inverter if AC loads), respectively.

Example System:

As an Example let's say we wish to power a Large Video Camera and Lighting System on a Camera Shoot in a remote location in Arizona for 24 hours. The Video Camera draws 5 watts, and the Lighting system is rated at 50 watts and we wish to run these two system 24 Hours straight with Solar Energy. What size Solar Power System would do this?

To design a Solar Power System begin with the Power required to drive your devices which will be expressed in terms of Watts.

Step One: Define your Electrical Load "Power" Requirements

List all of your electrical loads in terms of Watts. In our example the combined power load is 55 watts. This is the Power Rating of your electrical Load (in this example).

Step Two: Define your Daily Electrical Load "Energy" Requirements

The Next step is to define the Energy required by your loads per Day. Energy is Power over time so we'll put in our time measurements which in our example is 24 hours all day. To calculate your Energy demand multiply your Power by Time (55w x 24h) which is 1,320 watt-hours per Day.

Step Three: Determining your Solar Resource

Solar Energy which reaches the ground depends on specifically where you are on Earth. Near the equator has more raw solar resource than near the poles, but local "micro-climates" are also a factor. Solar Energy is therefore defined as equivalent "Peak-Hours" of sunlight which has a power value of

1,000 watts per Square Meter. In Arizona you may be at 5.5 Peak Hours of Sunlight. In the North-East you may be at 4.0 Peak Hours of Sunlight. In Portland, Oregon you may be at 3.5 Peak Hours. For your specific location look up your Peak Hour Solar Rating at NREL.

In our Example, we're in Arizona and we'll use 5.5 Peak hours. This is important as it tells you the average Energy you would expect from a given Power solar panel. In this location of 5.5 Peak-Hours a 100 watt Solar Panel would be expected to produce 550 Watt-Hours of Energy. Energy is the product of Power over Time.

Step Four: Sizing your Solar Panels

Now that you have your Solar Resource in Peak-Hours, in our example 5.5 Peak-Hours, you can calculate the Size of Solar Panel you'll need to produce the Energy required for your Load. In our example, the Energy Load is 1,320 Watt-Hours per Day.

To Calculate the Power rating of the Solar Panel to produce this energy divide the Energy Load by the Peak-Hours for your location. In our example (1,320 Watt-Hours/5.5 Hours) which comes to 240 Watts.

The Solar PV Panel required to produce an average energy of 1,320 Watt-Hours at a location with an average Peak-Hour rating of 5.5 Hours is 240 Watts.

Step Five: Define your Battery Requirements

To choose your Battery Size begin with defining a System Voltage. System Voltages for DC systems are usually 12, 24, or 48 VDC. The choice of Battery System Voltage by the designer depends on the Size of the System. The Larger the System, generally, the larger the System Voltage.

Note: The Power Loss in a circuit is a function of wire resistance and the Current (Amps) you're pushing through the wire. The Power Loss increases as the Square of the Current called the (I2R) rule which is Current Squared multiplied by the Resistance of the wire. Therefore, to avoid large currents (which have large losses in transmission), system designers choose higher voltages to transmit power through a system. Since "Power" is Voltage multiplied by Current (Amps) you can move the same Power with less transmission loss choosing a higher Voltage at lower Amperage.

Since most consumer electronics can be powered by 12 VDC power plug, and since our system is relatively low power (55 watts) we'll choose the 12 VDC format for this example.

In our example the Daily Energy Load was calculated at 1,320 watt-hours. The "Energy" available in a battery is calculated as the Voltage of the Battery multiplied by the Capacity of the Battery in Amp-hours (Ah). Energy (Watt-hours) = Volts x Ah

Step Six: Solve for Battery Capacity

For a Given Battery the Voltage is Selected, and the Size of the Battery is expressed as Amp-hours (Ah). Since our example is at 12 VDC our Daily Energy Load in Watt-Hours will equal the System Voltage multiplied by the Ah rating of the Battery we're solving for.

Solving for the Ah rating of our Battery for our example, we divide the Daily Energy Load (1,320 watt-hours) by the System Voltage(12 V) which comes to 110 Ah.

The Battery for our Daily Energy load will need to be at least 110 Ah in Size to run our Load for 24 straight hours.

Note: there is a difference between theoretical calculation and real world performance. The climate varies by the hour and daily energy is variable and difficult to exactly predict. At this point of the design process it's best to "De-rate" the system by Oversizing the Capacity of the Solar Panel and the Battery bank by at least 20% to compensate for variation.

The Emergency Sun system you'll be building in this book is rated at 1500 mAh @12VDC. This will charge your Smart Phone in 3-4 hours, or power your typical Tablet for 4-5 Hours

Chapter Three: Emergency Sun Parts and Procedures

The Emergency Sun is a portable Solar Power Supply which uses onboard solar photovoltaic cells to charge an onboard battery set composed of Twelve (12) Rechargeable AA batteries.

Rechargeable AA batteries have an individual Voltage which will range from 1.1 Volts usually near full discharge, or as high as 1.35 Volts each when fully charged. Therefore, the Voltage of Twelve batteries in series will vary between 13-16 Volts and this is exactly what you want.

Voltage runs downhill

A Solar Panel must produce a higher voltage than the Battery Bank you wish to charge. The energy stored in the Rechargeable AA batteries (about 1500 mAh) can be tapped at anytime through an output device either a 12 VDC universal cigarette lighter plug style, or USB through a plug-in 12 VDC/USB converter available at most electronics and hardware stores to power your electric loads anytime day or night. The driving "pressure" of this process is Voltage.

If you're using the Emergency Sun Kit you'll find a 5 watt solar photovoltaic panel. This is the primary charger for your system and provides the power to charge your internal battery bank and subsequently your loads producing a Voltage higher than your Batteries and electric load. The solar PV panel is rated a 5 watts which produces a Voltage under load of over 17 VDC.

The Emergency Sun is based on a 12 Volt system, and as such the solar panel produces 5 watts at peak sun.

Solar PV panels use individual solar cells each producing about 0.5 Volts. Solar panels therefore interconnect thirty six (36) solar cells in Series which increases voltage to 18 VDC. Solar PV panels are wired to this higher voltage because they are designed to charge batteries and a higher voltage than 12 VDC is required to charge. Remember, voltage travels "downhill" and flows from higher voltages to lower voltages and never the other way around.

Required Tools:

The Emergency Sun has electrical, mechanical, optical, thermal, and structural elements. To build a solar charger you'll need some fundamental tools which include

Soldering Iron
Solder - Electrical grade at least 40% flux
Wire Cutters
Wire Stripper
Soldering Holding Stand (optional)
Heat Gun (for heating heat shrink)
Small Metal Punch
Hand Drill with Philips Screw Bit
Measuring Tape
Scissors
Small Hammer

General Procedure

We begin by preparing your Solar Panel wires by cutting and stripping the wires. The Solar Panel wire you'll trim is used as your Output lead. Next, you'll build your electrical components which include your 12 VDC output plug, series connecting your battery trays, and finally soldering your control circuit all to be outlined in detail below.

After you've built your control circuit and tested for proper function, then you'll move into the mechanical side of this construction by building the Back Panels.

Once your Back Panels are put together by mounting industrial Velcro strips on the inside edges of the openings, then you'll mount the Control Circuit and build the mechanical structure. The final assembly involves mounting the Main Gasket and mounting the Back Panels to the Gasket producing a durable and rugged laminate integrating everything together.

The following Chapters will break down the specific procedure for each of these Steps as you build your Solar Power Supply. Also, there is a **Quick Guide** in the back of the book for easy reference.

Chapter Four: Preparing your Solar Panel

In this chapter we'll begin construction of your Emergency Sun by first preparing your Solar Panel for building. The specified Solar PV panel is rated at 5 watts and is built with an aluminum frame.

On the back side of the Solar PV panel there is a junction box which houses the Power Output leads which export the electrical power from the solar panel to your electrical load.

The Output leads, called a "pigtail" shows a casing housing a Black and Red wire, respectively. The Black wire is the Negative Lead with the Positive Lead being Red.

When you interconnect your Emergency Sun system everything will be wired in "Parallel" which is to say you'll wire Black to Black and Red to Red.

The Solar Panel wires which you cut and prepare will be used to build your Output Plug.

Step One: Preparing the Solar Panel

Turn the Solar Panel to the back side exposing the Junction Box and Pigtail.

The Solar Photovoltaic (PV) Panel is rated at 5 watts and produces direct current (DC).

On the Back of the panel you'll see the Junction Box affixed to the back where the output Wires come out.

The Black wire is Negative, and the Red wire is Positive.

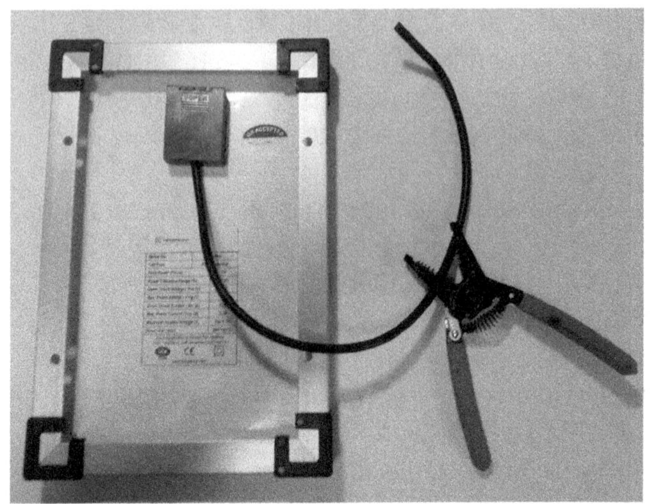

To prepare your wires you'll make two cuts.

First, a long section by cutting off a 9 inch length. This Length of wire will be used as your Output 12 VDC lead.

Second, cut the wires coming from the J. Box to a length 5" of wire casing. Next, carefully cut the casing back 2 inches from the end exposing the Negative (Black) and Positive (Red) wires.

Use great care when stripping the ends of the wires from the J-Box.

The outer casing is tough, and you need to strip the wires (intact) by just going deep enough to cut the outer casing, but spare the inner wires. (See the image above for reference).

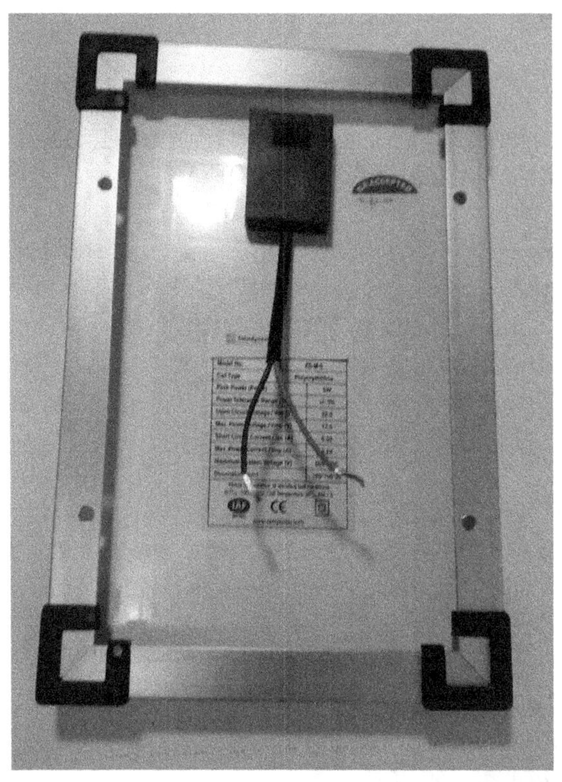

By hand you'll first strip the Outer Casing carefully so as not to nick the inner two wires.

The Black Negative wire and the Red Positive wire, once revealed should be stripped back at least an inch, or what ever length you'll need to set the wires in position and twist around one another until you're ready to solder.

Next, we'll begin constructing your 12 VDC Output Plug from the long wire you cut above.

Step Two - Building the 12 VDC Output Plug

Take your 9″ lead and wire strip each end with one end short (1/4″) to be soldered to your 12 VDC Plug, and the other end longer (1″).

There are two stripping operations, first you'll strip the Outer Casing off exposing the inner Red and Black wire - Careful not to nick those inner wires. Once the Casing is removed, strip each wire (Red, and Black, respectively) exposing the inner wire to which you can solder. (See photo below).

The 12 VDC Output plug connects your circuit (to be built below) to the outside world. The 12 VDC Plug is how you'll access your Emergency Sun's power for your device.

Be sure your solders are well done, and that you give them a light Tug to be sure they're mechanically sound.

You'll be Soldering the short wire stripped end of your 9″ lead to your 12 VDC Plug (after you've unscrewed the access connector).

Important Note:

The Center Clip of the Plug is Positive. The Edge connectors are Negative. Solder your Red (Positive) wire to the center. Solder your Black (Negative) wire to one of the edge clips.

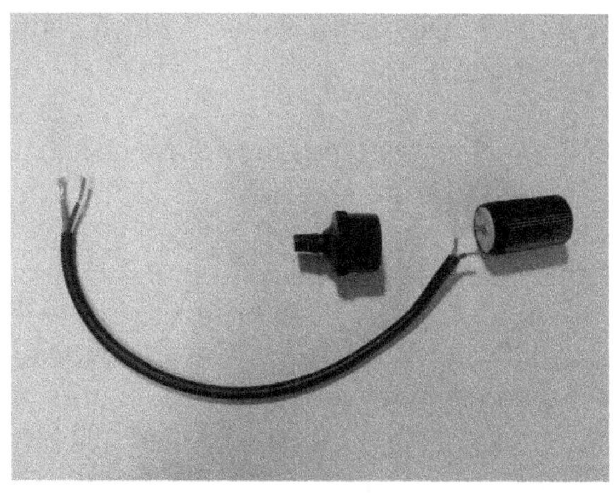

Pictured above is your 9" Output lead as you prepare to solder your 12 VDC Plug to one end of the wire you prepared by stripping the wire ends (be sure to slide on the end-cap to be ready to screw to the base of the 12 VDC Plug once you've soldered your leads).

Solder your 12 VDC Carefully (Red Positive Center, Black Negative Edge).

Be sure there are no strands of wire which can reach from one lead to another.

After you solder the plug leads (center Red, and edge Black) tug lightly at the solders to insure they make a solid connection. Next, thread on the end cap and Carefully screw on the Back Cap until you locked tightly together forming a strain relief for the plug. (See below)

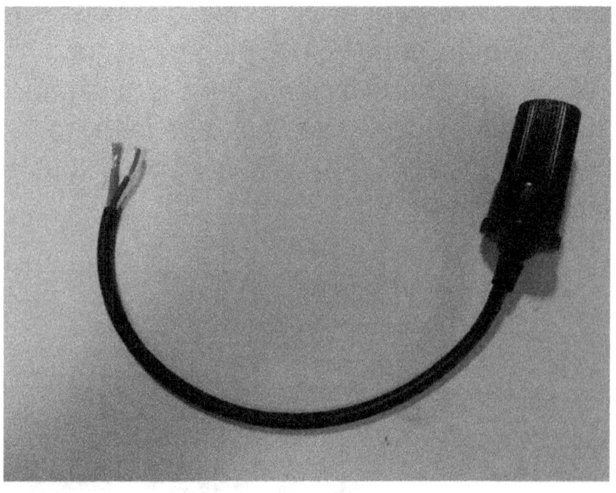

The Finished 12 VDC Plug assembly will now look like the image above.

Now, put this element aside and we'll continue with some mechanical work. Our next step is to construct our handle across the top of the Solar panel.

Step Three - Mounting the Handle

The next step is to attach your handle made from Polypropylene Webbing Material to the Solar PV Panel.

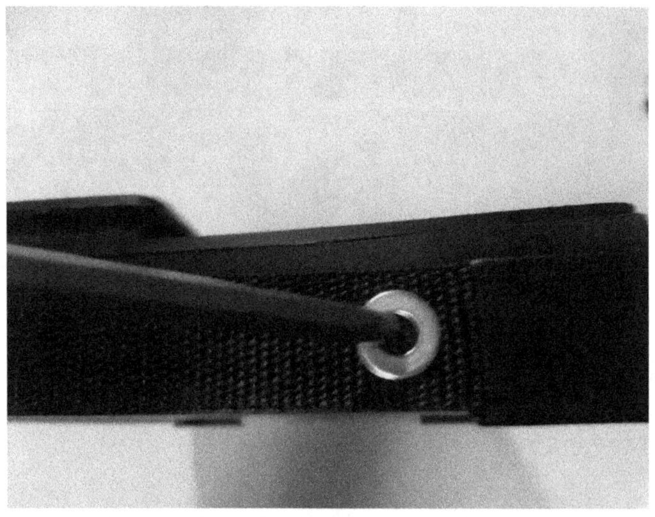

To build your handle you'll need your handle strap cut to size (6" of Webbing) in length and one (1") wide.

Use two (2) sets of screws (Lathe Screws 1/4 inch in length and flat head washers (1/2").

Position the Webbing material to One Side of the Top Edge and with the Washer 1/4" from the plastic corner piece. Using a Tap, firmly hammer a position mark.

Remove the Webbing, and push the screw through the washer and Webbing and reposition on the Top of the Frame.

Using your Hand Drill drive the screw through the Washer and Strapping and into the Aluminum Frame. Repeat on the other edge.

Chapter Five - Preparing Wires and Ready to Solder

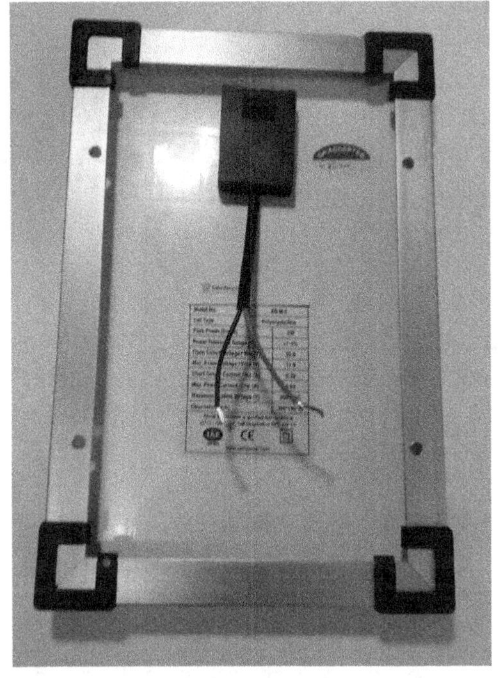

Step Four - Building the Battery Bank Series Connections

Now we begin building your battery bank Series connection.

In this step we'll combine three (3) separate battery trays which each hold four (4) AA Rechargeable batteries themselves in series.

Take Three of the Clip Connectors shown below and wire them in series.

First, strip the wire's ends back just a bit exposing more inner wire to work with, as they're easier to hand twist together, and cut two Small (0.5" in length) sections of Heat Shrink and thread them onto each Red wire. (After you solder you can slide the Heat Shrink up over the solder and blow with your Hot Air gun causing the Heat Shrink to shrink and seal your solder).

By hand, wire three Clips together in Series. Take one Clip. Note one end will have a Black (Negative) lead the other Red (Positive). Hand twist the Red wire from one clip to the Black wire of the next Clip to form a Series Connection.

Combining Three (3) clips in this way (with Heat Shrink threaded on before you twist the wires) you'll have a Series Connection between the clips.

You'll know you're correct with one Black end as the Negative Pole, and on the other end of the Series connection you'll have a Red (Positive) wire.

This is your series connection. Note the far right is the Black (Negative) battery pole, and on the far Left is the Red (Positive), between them is all in series.

In this way you're putting all of the battery trays, and batteries themselves, in a Series connection.

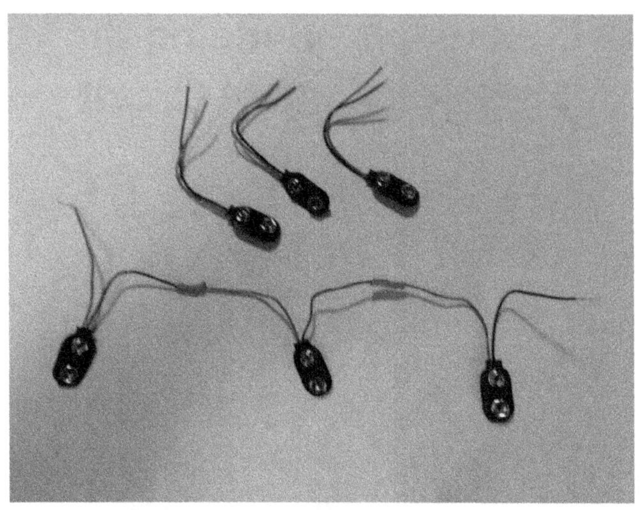

Finished Battery Tray Clips Series Connection

Short hand guide to this Step: By hand, twist together the Clip's wires in Series, with Heat-Shrink pre threaded on the wires.

Solder each twist and pull the Heat Shrink over the solder when cool.

Once the Heat Shrink is in place, use your Hot Air Gun to heat the Heat Shrink which will "Shrink" over your solder and provide great mechanical strength and electrical isolation.

You've now finished your Battery Tray Series Connection. Now, we're ready to build the electronic Control Circuit which will connect your Solar PV Panel with your Battery Bank with your Output Plug.

Chapter Six - Building the Electrical Control Circuit

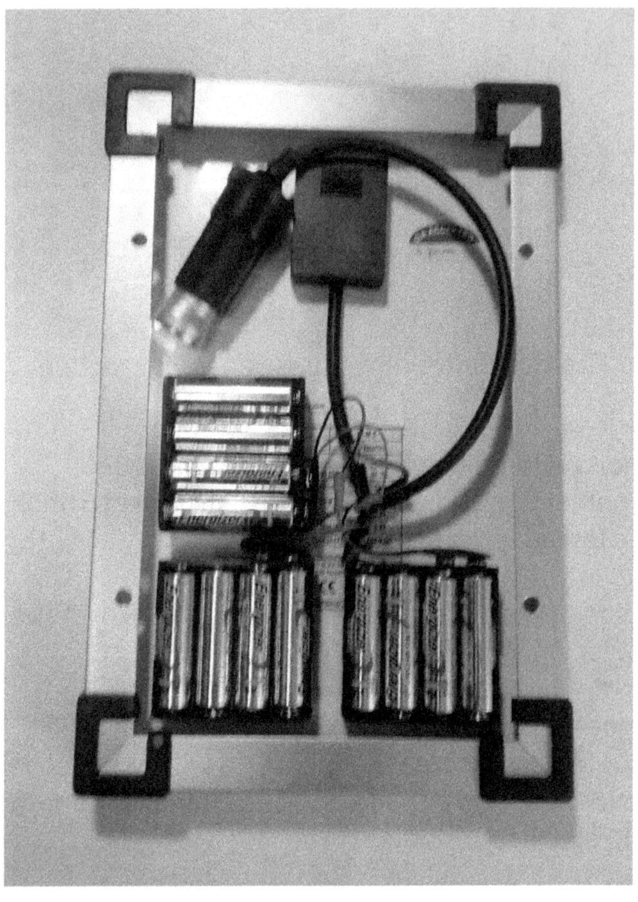

All battery charging systems require a Battery Management System (BMS) which serves to protect the batteries from three conditions: overcharging, over dis-charging, and back flow.

The Control Circuit

Solar PV panels produce a Direct Current having Positive and Negative leads, respectively. To charge a battery bank, in this case a 12 VDC battery bank, connect the solar PV panel and the battery bank in "Parallel" which is Black to Black, Red to Red.

Note: There are two types of wiring paths used separately or combined in electrical DC circuits: series, or parallel, respectively. A "Series" connection between solar cells, or individual battery cells increases voltage while the "current" produced does not change. In a "Parallel" connection the Current increases with the Voltage not changing.

The Control Circuit is also called a Battery Management System (BMS). As a "Charge Controller" a circuit is wired in-between a Solar PV panel and a Battery Bank to control three essential functions: Backflow- Overcharge or Over-Discharge.

A Charge Controller circuit is required in between a Solar Panel and a Battery because wiring a Solar PV panel correctly in Parallel (Positive to Positive, Negative to Negative) with a battery works fine when exposed to the sun. The Solar PV panel produces a higher voltage than the battery bank and "pushes" charge into the battery by the pressure of the higher voltage.

If the Solar PV panel is taken out of the sun, then "charge" accumulated in the battery begins to

"Reverse Flow." Without protection and wired directly the solar PV panel looks like a Resistor when out of the sun and the battery begins to discharge - defeating the purpose of charging.

The answer to this issue and one of the primary functions of a Charge Controller is to prevent this "Reverse Flow" by using a "Blocking Diode."

Step Five - Building the Control Circuit

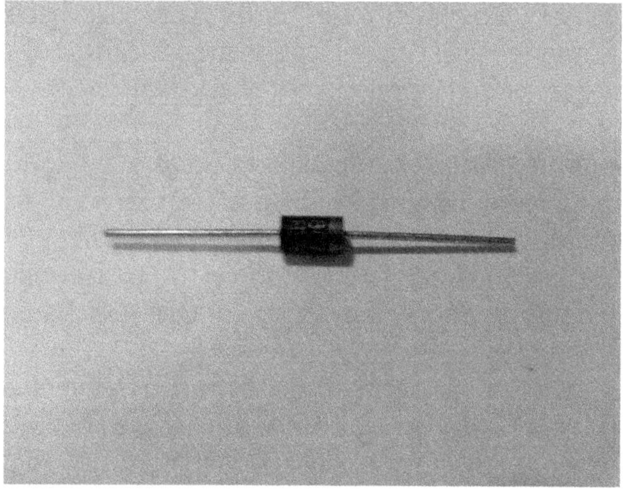

Solder your Blocking Diode (Reverse Bias at least 100 Volts) such that the Band is facing away from your Solar PV Panel Pigtail Red (Positive) lead. Heat-Shrink after Soldering. The Blocking Diode will have a Cathode band which is on the right hand side in this picture (barley visible). This is the negative side of the Diode.

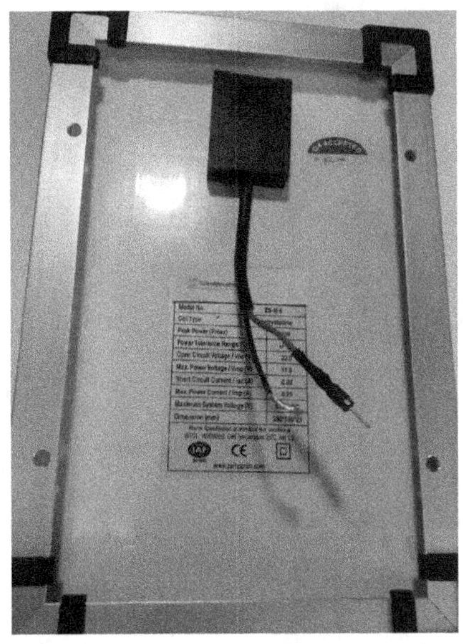

Note: Be sure the Diode is pointing the right direction before you solder and heat shrink. Cathode band out. Your Solder and Heat-Shrink will look like the image above.

Step Six - Soldering the Negative Leads (Parallel)

The next step is to start pulling your whole circuit together by soldering together all of your Negative leads.

Combine your Negative Solar Panel Output wire (Black), with the Negative (Black) wire from your Battery Series Connection, with the Negative (Black) wire of your 12 VDC Output Plug.

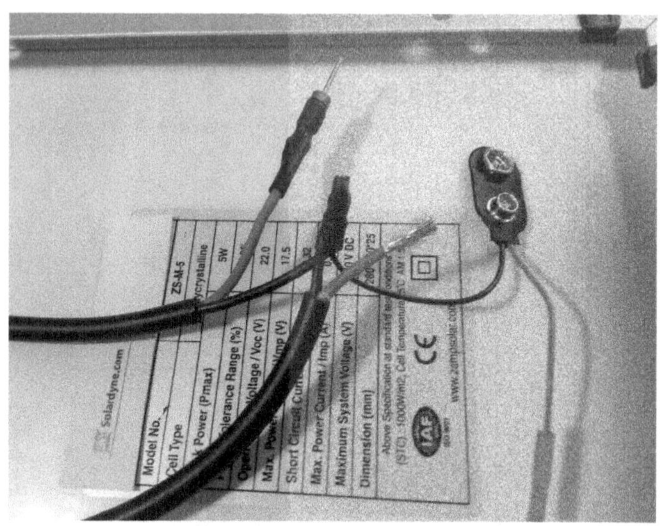

Pictured above are your three Negative (Black) leads being hand twisted together then Soldered, and Heat-Shrink after you've tugged on all the leads to be sure the Solder is strong. After Soldering, then Heat-Shrink.

Step Seven - Soldering the Positive Leads (Parallel)

Combine all of the Red (Positive) leads together. Specifically you'll hand wire in preparation for soldering all Three Positive wires: the Red (Positive) lead to which you soldered the Diode (now you'll solder the other end), the Red (Positive) lead from the Battery Series Connection Clips, and the Red (Positive) lead to the 12 VDC Output Plug. (See Below)

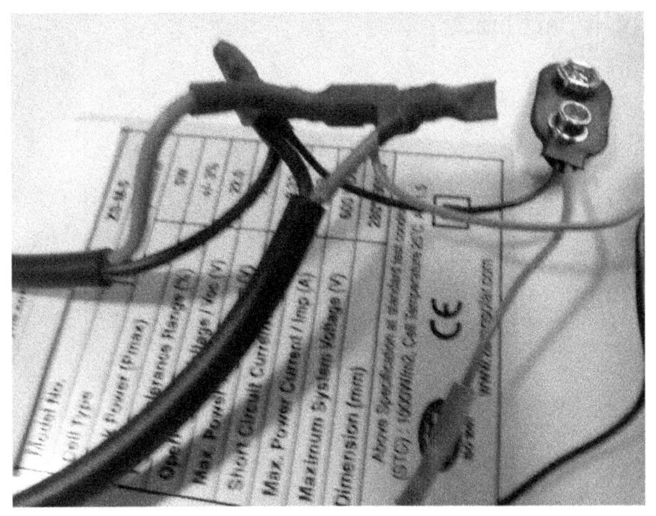

Shown above, you've soldered and heat-shrink wrapped first the Black (Negative) leads. Then you'll soldered and heat shrink wrapped the Red (Positive) leads. You've now finished building your Control circuit. Remember, all Black leads have been soldered together, and separately, all Red leads have been soldered together.

Step Eight - Preparing the Battery Trays

Each Battery Tray will hold four (4) AA Rechargeable Batteries. Each Battery Tray internally wires these batteries in Series. We've further connected these Three (3) Battery Trays in Series which places all twelve (12) Rechargeable AA batteries in Series.

Next, we'll connect the Battery Trays to our individual Clips.

To mount the Battery Trays to the Back of the Solar Panel, take your Master Gasket and punch out the Internal Rectangular Gaskets. There are four (4) internal Rectangular gaskets provided.

To mount the Battery Trays to the Back of the Solar Panel take Two (2) of the Rectangular gaskets and cut them in half.

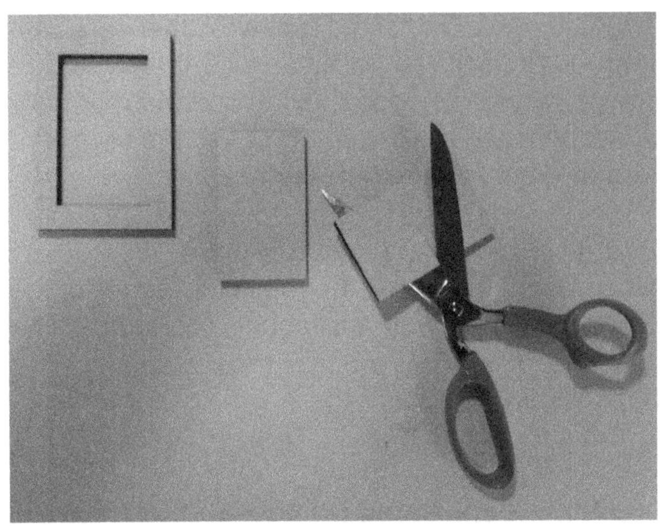

Each half has a Pressure Sensitive Adhesive (PSA) on each side. Remove one side of the White Wax Paper and stick it to the Back of each of the three (3) Battery Trays.

This Rectangular Pressure Sensitive Adhesive (PSA) surrounds a thick thermal and shock absorbing material. This provides a shock absorbing thermal barrier between the Battery Trays and the back of the Solar Panel to which their affixed.

Note: in extreme temperatures such as the Outback this approach keeps the thermal temperature of the Batteries separate from the high temperature of the Solar Panel itself. This protects the battery thermally, and mechanically.

Step Nine - Connecting the Battery Trays into a Series connected Battery Bank.

Each Battery Tray has terminals on top which fit a 9 Volt Clip you wired in Series above.

Snap Connect Each Battery Tray (there will be three in total) using the snap connectors to your Control Circuit. There is only one way they will snap onto the Battery Trays - so no errors are possible.

Mount first the Bottom Right Battery Tray which will have the Black (Negative) wire soldered to the Negative leads.

Mount this first Battery Tray in the Lower Right corner as shown below.

Next, repeat the procedure Clipping the next Battery Tray to the next Clip, and positioning now in the Lower Left corner - in mirror position as the first.

Repeat with the Third Battery Tray rotated 90 degrees to keep the Clip connector towards the center, connecting by Clip and positioning Above the lower Left Tray.

Allow 1/4" separation between the Battery Trays evenly. See image below.

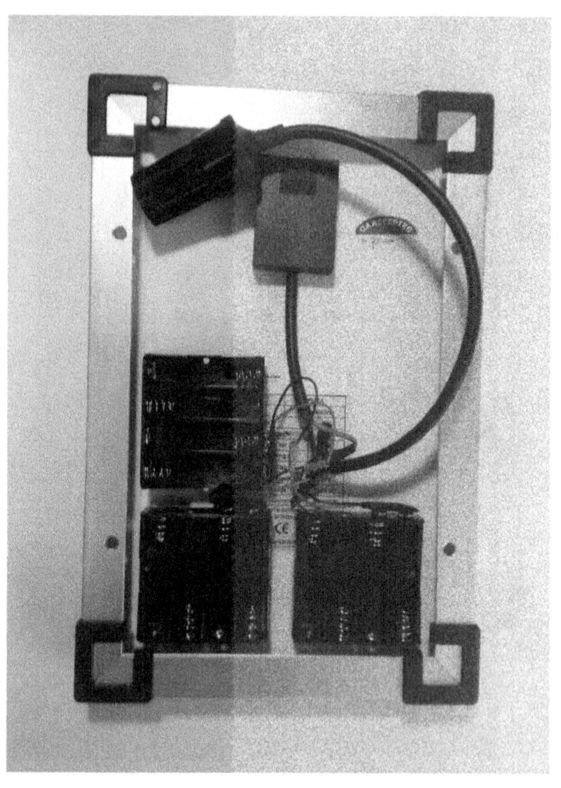

Note: if you're not sure of your Battery Tray positions, then put the Back Main Panel with the hole for the batteries over the Frame and lined up being sure there is enough room to comfortably put your batteries in the Battery Trays through the Back Frame.

All Three (3) Battery Trays now mounted we're ready to electrically test our circuit before we mount anything further. If there is an issue we want to find it now before mounting become permanent.

Chapter Seven - Electrical Testing

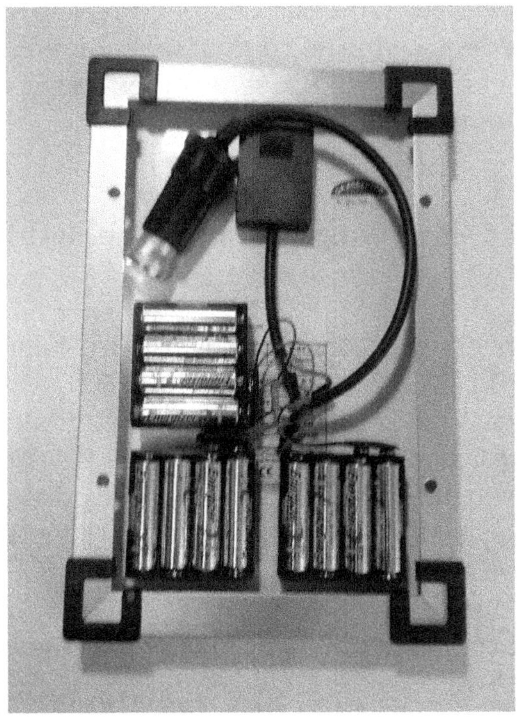

Testing is a vital part of assembling your Emergency Sun solar power system. Testing your electrical circuit using a Multimeter is important by verifying everything is working correctly as you build, mount and use your circuit.

The important elements to test electrically are the Solar Panel as your primary charging source of power with the Battery Charging functions.

Testing your Solar Panel

The Open Circuit Voltage is measure by placing the Solar active surface in the Sun and placing your Negative Lead on the Black output wire, and the Positive Lead on the Red output wire. Turning your Multimeter to DC Voltage settings you'll see a measurement near 18 VDC.

Step Ten - Measuring and Testing the Electrical Circuit

You can test your Control Circuit using a Multimeter, or your 12 VDC/USB converter. The USB converter usually has a Blue LED light which goes on when you insert it into a 12 Volt Plug. If there is sufficient Voltage to turn on the light, then you're good to go in using your Power Plant as a Blue LED light shows you have sufficient Voltage in your Battery Bank.

A more comprehensive way to test your Emergency Sun is to use a proper Multimeter.

Select "DC Voltage" on your Multimeter. Use your Black probe for Negative and your Red Probe for Positive.

Test One - Testing Battery Series Connection

Once the Rechargeable AA batteries are loaded the circuit is live. To measure the Voltage of the Battery

Bank take your 12 VDC Output Plug and Measure the Voltage from the Plug.

To Measure the Voltage, Carefully place your Positive RED lead down the center of the Plug and contact the Center Contact. Place the Negative BLACK lead to any point on the inside cylinder which is all Negative.

Your Multimeter should read above 12 VDC and as high as 16 VDC.

Note: If you don't measure at least 12 VDC then you have an issue in your Circuit. Immediately remove the batteries. Inspect your circuit for the physical problem and rebuild. Repeat Test until you see a steady Voltage of at least 12 VDC.

Test Two - Testing the Solar Charging Function

To test if your Circuit is charging the batteries properly, load your Emergency Sun with twelve (12) Rechargeable AA batteries into the Battery Trays. Set your Multimeter to DC Voltage and Measure the Voltage of the Battery Bank using the same method as above.

Place the Positive RED lead down the center of the Plug to contact the center terminal, and and place the Negative BLACK lead on the inside side surface of the Plug to complete the circuit and your Multimeter should now read a steady Voltage of 12-16 VDC.

Make a note of this Voltage _______.

Take the Emergency Sun outdoors and place the Solar Panel surface in the sun for a period of time - say 5 minutes.

After 5 minutes in the Sun take another measurement of the Battery Voltage and make of note of this Voltage ________.

If the Second Voltage is Higher than the First Voltage you measured before the test, then your Solar Panel is Charging Properly.

Note: if the Voltage has not increased, then you have an issue with your Control Circuit. Immediately remove the batteries. Inspect, and rebuild the Control Circuit. Repeat Test until you record an Increase in Voltage after Charging.

Chapter Eight - Mounting the Circuit

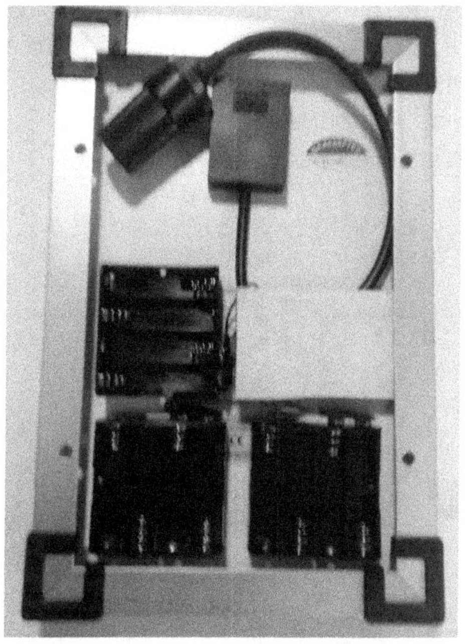

In this Chapter we'll focus on the mechanical mounting of your Control Circuit.

Step Eleven - Mounting the Control Circuit

Using two (2) of your Rectangular Sticky Pads used to mount your Battery Trays you'll use one as a bottom Pad which you'll affix to the Back of the Solar Panel Above the Lower Right Battery Tray, and in the Space Hidden by your Back Panel.

The Rectangular Sticky Pads are sized to fit above the Right Lower Battery Bank. Remove one side of the White Wax Paper and place one Sticky Pad directly on the Back of the Solar Panel to provide a base to which you'll Stick Down the Control Circuit. Remove the White Wax paper on the Top surface and push down your electronics onto the Sticky surface to secure. (See below)

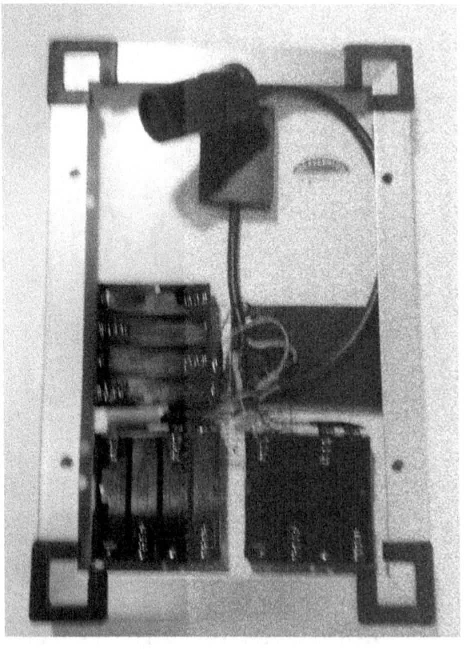

Now that the base Sticky Rectangle is placed under the electronics (shown above), and you've pushed down the electronics into the Sticky Pad to secure, then take your Second Pad and remove one of the White Wax paper surfaces.

Affix the second Sticky Rectangle as the "top of the sandwich" and press down firmly. Leave the upper Wax Paper affixed and press down to be sure you're covering as much surface as possible between the two sticky surfaces.

A firm "Sandwich" provides the "strain relief" on your 12 VDC Output Plug by keeping the base near your electronics snuggly pressed between the two Sticky pads keeping the 12 VDC output plug secure.

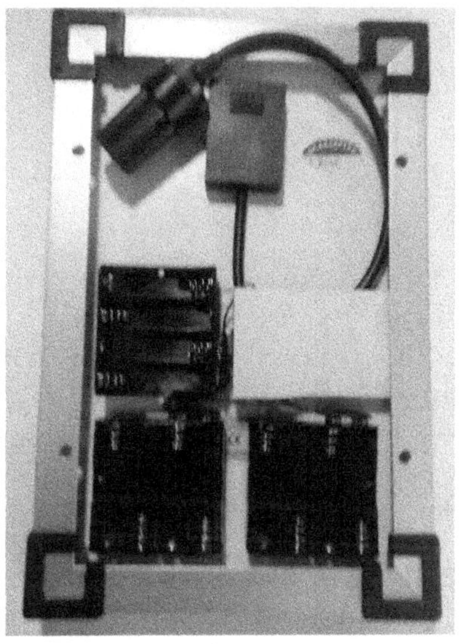

With this step you've mounted your internal Control Circuit which connects the Power Input, with the Battery Storage, and the Power Output Plug bringing all of the electronics together.

Next, we'll focus on building the Back Panels and mounting these Panels to the Solar Panel.

Chapter Nine: Building your Back Panels

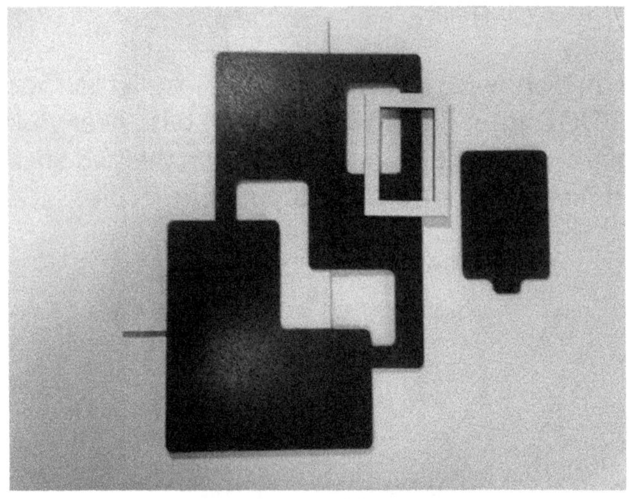

The 5 watt Solar PV Panel is constructed with a study aluminum frame which will act as the framing and main structural support for your Emergency Sun system.

Industrial Velcro (3/4" wide) is used to connect the Back Panel Covers with the Main Back Panel.

Assemble the Back Panels by Measuring and Cutting the high strength Velcro which surrounds the openings for the Battery Trays (the Big L Shape), and the Upper Panel for accessing your 12 VDC Output Plug, you'll affix the "Fuzzy"side on the Main Base Panel as a collar which surrounds the openings.

Step Twelve - Building the Velcro Seals for the Large Back Panel

The Main Back Panel (shown below) is the main Back Cover plate. This will be affixed with the Main Gasket to the Solar Panel later in the procedure.

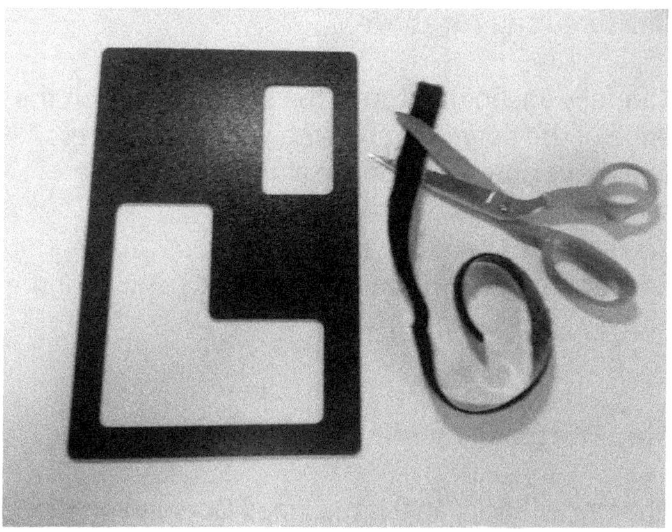

Measuring the Velcro

Velcro has two parts, a Fuzzy bottom surface and a Hooked Top surface, respectively. Start with your Fuzzy bottom Velcro and measure out each Side of your Battery Access "L" space as shown above. Measure and Cut each side strip.

This Velcro is 3/4" wide. You'll stick the Fuzzy Velcro layer as a "Frame" around the "L" opening seen above.

Attaching the Velcro

Velcro has a Pressure Sensitive Adhesive (PSA) protected by a plastic film.

Remove the film exposing the Pressure Adhesive and press each length of Velcro Fuzzy sections you measured and cut above.

Carefully anchor the top of the sticky surface on the edge of the openings and with your fingers press and move down the surface along the opening securing the Velcro to the Rim.

Repeat until you Surround the large Opening for the Battery Trays.

Step Thirteen - Building the Upper Back Panel

The Upper Back Panel is the Small Opening on the upper right of the Back Panel.

This is the access panel to your 12 VDC Output Plug.Because the 12 VDC Plug is thicker than the Solar Panel, we'll add a Small Picture Frame Gasket which are included in your kit.

The Small Picture Frame Gasket lifts the Upper Back Panel up a 1/4" to provide more room for the 12 VDC Output Plug.

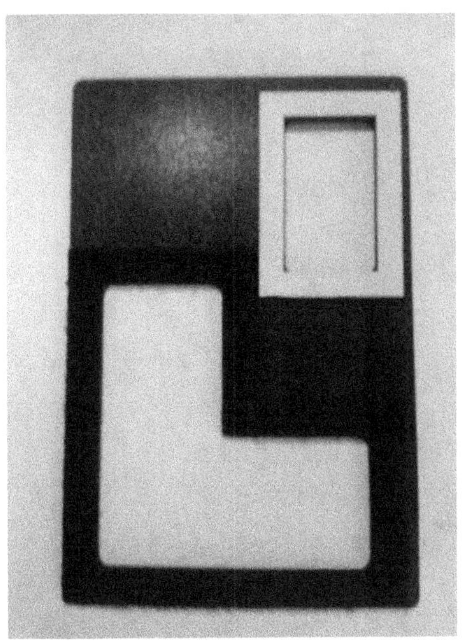

Take the Small Picture Frame Gasket (shown above) and remove one side of the White Wax Paper to expose the Sticky side. Place this side around the Small Upper Panel opening (as shown above) to add some height to the Upper Back Panel opening.

Take your 3/4" Velcro supplied and Trim into a length of 1/2" diameter (which is to say, trim 1/4" of the width of the Velcro Fuzzy Side.

Measure your Trimmed Velcro (now 1/2" width) and Cut to size to fit the Small Picture Frame Gasket. Remove the White Wax paper exposing the Sticky surface, and affix the trimmed "Fuzzy" Velcro. (See below)

Your Back Panel (shown above) is now complete. Now, we're ready for the Back Panel Covers.

Step Fourteen - Building the Back Panel Covers

Now that the Back Panel is complete let's move on to the Back Panel Covers. There are two (2), Back Panel Covers: one to Cover the Battery Bank opening (the Large Space), and the Second to cover the Output Plug Upper Panel.

Each Cover Plate (shown below right) will be constructed using the "Hook" side of the Velcro attached to the Inside Surface to Mirror the Velcro attached to the Main Back Panel.

Take your Two Upper Panel Covers and turn over to expose the Inside Surface. You'll affix the "Hook" side of the Velcro to these inside surfaces mirroring the "Fuzzy" Velcro of the Base Plate.

Once you've attached the "Hook" Velcro to the inside surface of the Back Panel Covers (shown above) then you're ready for the Final Assembly.

Chapter Ten - Mounting the Back Panel and Final Assembly

Congratulations, you're almost there.

The Next Step is to mount the Main Gasket which laminates your Back Panel to your Solar Panel to complete the Assembly.

Step Fifteen - Mounting the Main Gasket

The Back Panel, which you've built above is now ready to mount on the Solar Panel Frame. You've tested your circuit by loading the Rechargeable AA batteries and Measured your Voltage and now you're ready to seal up the Back Panel.

Using the Large Main Gasket remove one surface of White Wax Paper to expose the sticky Pressure Sensitive Adhesive (PSA). Carefully line up the Gasket with Solar Panel Frame (on the backside of the Solar Panel) and Press Firmly.

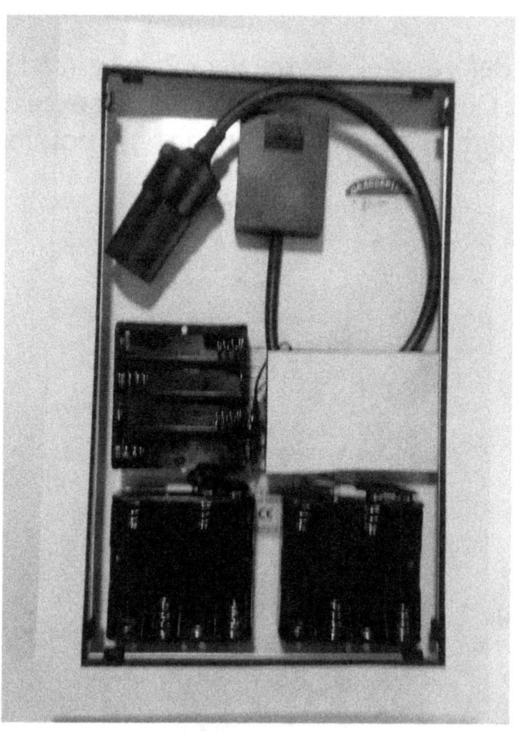

Shown above you can see the Main Gasket is now Mounted on the Back Frame of the Solar Panel.

Step Sixteen - Mounting the Back Panel

We're now beginning the Final Assembly with the mounting of the Back Panel to the Main Gasket which affixes the Main Back Panel with the Solar Panel Aluminum Frame.

Remove the protective White Wax Paper of the Main Gasket exposing the Sticky surface and Carefully position your Main Back Panel for a Press fit.

Note: Before you Press the Back Panel thread your 12 VDC Output Plug through the Upper Access panel so the 12 VDC plug is in the right position.

Shown above is your Back Panel firmly pressed onto the Main Gasket with each Panel Open. For a final test of the system, load your twelve (12) Rechargeable AA batteries and Plug in your USB converter into the 12 VDC output plug. You should see the blue LED light on the converter light up.

Step Seventeen - Final Assembly - Close Up the Back Panel Covers

There you have it! You've done it. Shown below is an image of the Back Side of the Emergency Sun with the Back Panels closed.

Congratulations. You've just built your own Personal Power Plant.

Chapter Eleven - Using your Emergency Sun

To use your Emergency Sun open your main Back Panel and load up twelve (12) Rechargeable AA batteries into the Battery Trays - taking care to load carefully and under the right orientation (See the inside of the Battery Trays for images).

Once the Twelve (12) Rechargeable AA Batteries are loaded after opening your Large Back Panel, put your Emergency Sun into the sun to charge with the Solar Panel side facing the sky.

After the equivalent of four (4) peak sun hours your internal batteries will be charged.

To power your Smart Phones or other electronics anytime day or night, simply open the Upper Back Panel exposing the 12 VDC Output Plug. You can plug in your 12 VDC adaptors and use directly as a 12 VDC plug, or plug in your USB converter and power your electronics via USB.

Enjoy your Portable Solar Power Plant and have great use of this versatile power supply. Outdoor users, RVs, Campers, First Responders, and Emergency use. Everyone can benefit from a

Rugged, Portable, Non-Toxic, and powerful Solar Power Plant.

Emergency Sun Night Time Use

Have Ready to for Natural Disasters

Chapter Twelve - Quick Guide

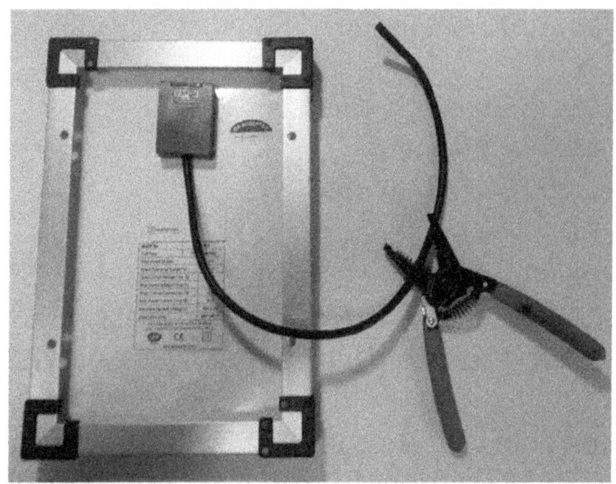

Suggested Tools:

You'll need some basic tools to construct this Emergency Sun Solar Power Supply they include:

Soldering Iron
Solder (at least 40% flux)
Wire Cutters
Wire Strippers
Heat Shrink Material (1/8" and 1/4")
Heat Gun
Hand Drill
Sharp Edge
Measuring Tape
Scissors
Metal Punch
Small Hammer

Parts List:

(1) Solar Photovoltaic Panel Rated at 5 Watts (11″ x 7.5″ x 2″)
(1) Blocking Diode (100 V bias)
(1) 12 VDC Output Plug
(3) Battery Trays (AA with Clip)
(3) 9 VDC Clips with Leads
(1) Large Picture Frame Gasket (7.5″ x 11″ x 0.75″)
(4) Rectangular (2″ x 4″) Double Stick Gasket
(1) Small Rectangular Picture Frame Gasket (2.5″ x 3.5″)
(1) Three Foot (3′) length of 3/4″ wide Velcro both sides with Pressure Sensitive Adhesive (PSA)
(1) Large Back Panel Base (7.5″ x 11″ with openings for Batteries and Output Plug
(1) Large Back Panel Cover
(1) Small Back Panel Access Cover

All of these parts are available in Kit form: the Emergency Sun Kit.

You can also source these parts independently and source everything yourself. Each reader is granted a Personal License to use this design for private use.

Use this book as a guide for parts and procedures to build your own portable Solar Power Supply.

To Source the complete Emergency Sun Kit visit Solardyne.com

Procedure Quick Guide:

About the Author

Christopher (Toby) Kinkaid

Christopher (Toby) Kinkaid, from Portland, Oregon is the founder of **Solardyne.com**, **SolarQuote.com**, and **AlgaeToday.com**, and has worked in the clean energy field for over three decades.

Mr. Kinkaid is the inventor of the "**Helyx**" Vertical Axis Wind Generator, the "**Mariposa**" Non-imaging solar concentrator PV module (continuous operation at Sandia National Laboratory since 1994), the **Solar Demultiplexer** optical solar concentrating lens (Dr. James/Sandia National Laboratory 1991), and the inventor of the original "**Solar Power Pack**" (Mother Earth News, "**Littlest Utility**" June/July, 2001).

Mr. Kinkaid has been an official lecturer and presenter on clean energy technology around the world including APEC, Bangkok, Thailand, 2003,

"Energy Solutions World", Tokyo, Japan, 2003, the International Biomass Conference (IBC), 2010, Minneapolis, MN, and the Algal Biomass Organization (ABO) Conference, 2010, Phoenix, AZ.

Mr. Kinkaid has appeared in TV interviews on KOIN TV, KGW TV, FOX 12 NEWS and "Sustainable Today" produced in Oregon.

Mr. Kinkaid has served on the board of directors for the National Hydrogen Association, in Washington D.C., 1993, and the Japanese Satellite Communications Company (JCNET), Fukuoka, Japan, 1994, Oregon Wind Corporation 2003-2007, Algaedyne Corporation (2008-2011), currently serving as CEO of Solardyne in Portland, Oregon.

Christopher (Toby) Kinkaid is based in Portland, Oregon, and continues his work in clean energy technology and applications.

www.ingramcontent.com/pod-product-compliance
Lightning Source LLC
Chambersburg PA
CBHW060408190526
45169CB00002B/799

* 9 7 8 1 5 2 3 6 1 6 7 7 0 *